Food 225

耕种鱼食

Fish Farming Their Food

Gunter Pauli

[比] 冈特·鲍利 著

[哥伦] 凯瑟琳娜·巴赫 绘

颜莹莹 译

上海远东出版社

丛书编委会

主　任：贾　峰

副主任：何家振　闫世东　郑立明

委　员：李原原　祝真旭　牛玲娟　梁雅丽　任泽林

　　　　王　岢　陈　卫　郑循如　吴建民　彭　勇

　　　　王梦雨　戴　虹　靳增江　孟　蝶　崔晓晓

特别感谢以下热心人士对童书工作的支持：

匡志强　方　芳　宋小华　解　东　厉　云　李　婧

刘　丹　熊彩虹　罗淑怡　旷　婉　杨　荣　刘学振

何圣霖　王必斗　潘林平　熊志强　廖清州　谭燕宁

王　征　白　纯　张林霞　寿颖慧　罗　佳　傅　俊

胡海朋　白永喆　韦小宏　李　杰　欧　亮

目录

Contents

一条小热带鱼正紧张地游来游去。一只海胆好奇地问：

　　"你在害怕什么？"

　　"我并不害怕什么事或什么人，但我随时准备保卫我的农场。"

A damselfish is swimming around nervously. A sea urchin wonders what is going on and asks:

"Are you afraid of something?"

"I am not afraid of anything or anyone, but I am always ready to defend my farm."

一条小热带鱼正紧张地游来游去

A damselfish is swimming around nervously

weeding out all the algae

"你的农场？我不知道你在耕种。我还以为你只是想保持你的房间干净呢。"

"我正忙着清除所有我不想要的藻类。"

"清除藻类？听起来工作量很大。周围有成千上万细小的藻类，你的余生将会非常忙碌。"

"Your farm? I didn't know that you were farming. I thought you were just keeping your area clean."

"I am busy weeding out all the algae that I do not want."

"Weeding algae? That sounds like a lot of work. With thousands of tiny algae around, you are going to be kept very busy for the rest of your life."

"看，如果我不清除这些讨厌的藻类，我最喜欢的那一种就不会生长。"

"最喜欢的藻类？那么，你是在告诉我，你只保留一种藻类，把其他的都清除？这听起来像是陆地上的人，他们只种植一种作物，然后喷洒化学品来清除其他植物。"

"嗯，我不用化学品。但这确实意味着我一直都很忙。"

"Look, if I do not remove these unwanted algae, my favourite one will not thrive."

"A favourite algae? So, you are telling me that you are only keeping one type of algae and removing all the rest? That sounds like people on land, who only farm one crop and spray chemicals to remove all other plants."

"Well, I don't use chemicals. But that does mean I'm kept busy all the time."

只保留一种藻类

Only keeping one type of algae

我的胃很娇弱

I have a very delicate tummy

"你为什么只吃一种藻类？我可不像你那么挑食，为了填饱肚子，我什么都吃。"

"你看，我的胃很娇弱。我不能很好地消化纤维，所以我得打理花园，只吃那些对我有好处的。"

"这么说，你们就像那些在陆地上种蘑菇的蚂蚁和白蚁一样？"海胆问道。

"Why do you only eat one type? I am not as picky as you are, and will eat anything and everything to fill my stomach and still my hunger."

"I have a very delicate tummy, you see. I can't digest fibres very well, so I need to tend to my garden, and only eat the ones that are good for me."

"So you are like the ants and termites, that grow mushrooms on land?" Sea Urchin asks.

"对。他们在地下建立自己的安全世界，而我在大海里耕种。"

"如果你知道我有多喜欢红藻，你一定会很紧张。我很想把它当作午餐……"

"Yip. They create their secure world underground – while I farm in the open sea."

"You must be nervous having me around, knowing how much I love red algae. I am very tempted to make a lunch of it…"

我在大海里耕种

I farm in the open sea

等着鱼死网破吧！

Expect a head butt!

"呃，你是我热带鱼花园里的彼得兔，这已不是什么秘密了。但你最好离我的地盘远点，否则我会狠狠揍你一顿。如果这还不能使你打消念头，那就等着鱼死网破吧！"

　　"看来你很有决心，热带鱼农夫，而且还很勇敢，即使你个头很小。"

"Oh, it is no secret that you are the Peter Rabbit of my damselfish garden. But you had better stay off from my turf, or you can expect a hard hit from my snout. And if that doesn't deter you, expect a head butt!"

"Sounds like you are very determined, Farmer Damselfish, and very brave for someone your tiny size."

"是的。即使我只有10厘米长，10克重，我也会奋起攻击所有威胁到我们食物的人。"

"你把藻类花园建在海葵旁边是非常明智的。他们的刺扎得很痛，我根本不想靠近他们。"

"Yip. Even though I am only 10 centimetres long and weigh only 10 grams, I will not shy away from attacking anyone threatening our food."
"Very wise to make your algae gardens right next to these anemones. Their sting is so painful that I don't want to come anywhere near them."

他们的刺扎得很痛

Their sting is so painful

我甚至还追赶潜水员

I even go after divers

"我需要所有能得到的保护。但我要警告你：如果你靠得太近，我就会对你穷追不舍。我甚至还追赶潜水员。"

"即使这意味着要和比你重一千倍的对手较量？我一直认为最好斗的鱼是鲨鱼和梭鱼。"

"I need all the protection I can get. But be warned: Come too close and I will come after you. I even go after divers."
"Even if it means taking on someone a thousand times heavier than you? And I always thought the most aggressive fish are sharks and barracudas."

"它们捕食生肉，而我和我的伙伴海葵则保护我们的食物。这就是为什么我不怕任何人的原因——不管他们有多大的牙齿或爪子。"
　　……这仅仅是开始！……

"They are predators that kill for raw meat whereas my partners, the anemones, and I defend our food. That's why I am not afraid of anyone – whatever the size of their teeth or hands."

... AND IT HAS ONLY JUST BEGUN!...

……这仅仅是开始！……

… AND IT HAS ONLY JUST BEGUN! …

Did You Know ?

你知道吗？

Scientists identified damselfish, in 320 different territories, that were found to be farming 15 types of algae on coral reefs, from Egypt, Kenya, Mauritius, the Maldives, Thailand, and the Okinawa, to the Great Barrier Reef.

从埃及、肯尼亚、毛里求斯、马尔代夫、泰国、冲绳岛到大堡礁，科学家在320个不同地区发现小热带鱼以珊瑚礁上的15种藻类为食。

Damselfish are the most aggressive fish on the reef. The White Spotted Devil (Plectroglyphidodon lacrymatus) is the most aggressive of all, fiercely defending its algae growing on lime stone patches.

小热带鱼是珊瑚礁上最具攻击性的鱼类。白斑魔鬼（Plectroglyphidodon lacrymatus）是其中最具攻击性的一种，它会极力保护生长在石灰石上的藻类。

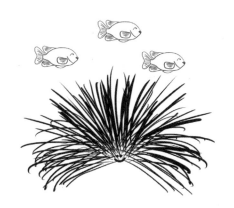

The symbiosis between damselfish and sea anemones is an exchange of guarding duties with protection by use of stinging cells. The anemones will sting all except the damselfish that share their territory.

小热带鱼与海葵的共生关系，是它们利用带刺细胞互相保护。除了与海葵共享领地的小热带鱼，海葵会用刺蜇其他所有生物。

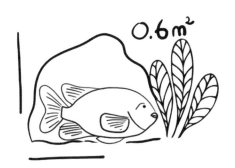

Most damselfish spend the majority of their life within a territory of about 0.6 m² in size. Their territories provide for their dietary needs, and contain spawning sites, as well as hiding spots.

大多数小热带鱼大部分时间都生活在约 0.6 平方米大小的区域内。它们的领地为它们提供饮食需要，包括产卵地点和隐藏地点。

There are 300 different types of damselfish. The juvenile damselfish typically has different colouring to that of the adults.

小热带鱼有 300 种不同种类。幼鱼的颜色通常与成年鱼的颜色不同。

The male Cortez damselfish takes care of the eggs. He may devour up to 25% of the clutch of eggs, so females will visit repeatedly to add eggs to his clutch, as small clutches are more likely to be consumed because they do not warrant the cost of parental care.

雄性科特斯小热带鱼负责照看鱼卵。它可能会吃掉多达 25% 的卵，所以雌鱼会反复来添加鱼卵，小的卵由于不值得被照料，会更有可能被吃掉。

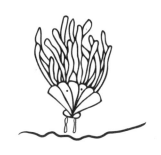

海胆身体下面有成排成对的管足，末端是吸盘，便于它们在海底移动，并在移动中捕猎，还能附着在岩石上抵抗巨浪。

Urchins have rows of paired tube feet underneath their body, ending in suckers that facilitate movement on the ocean floor, permit hunting while on the move, and adhesion to rocks to withstand heavy waves.

海胆有一种特殊的口腔，叫作"亚里斯多德的灯笼"，它有五颗锋利的牙齿，可以在岩石上钻洞。红海胆在野外可以存活 200 年。

Sea urchins have a special type of mouth, called "Aristotle's lantern", equipped with five sharp teeth that are able to drill a hole in the rock. Red sea urchins can survive for up to 200 years in the wild.

Think about It

Would you be able to live in one tiny spot all your life?

你能在一个小地方生活一辈子吗?

Is it a good idea to farm only one crop?

只种植一种作物是个好主意吗?

When defending your turf, would you take on someone bigger than you?

在捍卫自己的地盘时,你会和比你强大的人较量吗?

Who were the first farmers on Earth?

最早在地球上种植的是什么人?

How many people know that there are fish farming their own food? How many people know that ants and termites were already farming 300 million years ago? Make a list of all the animals that you can find that are cultivators of their food crops. Now share your list with friends and family members. Carefully observe how surprised they are to learn of this. Share more information on a few of the species on your list, and specifically on the damselfish, and how efficiently they are spending their entire lives in a territory just 0.6 square metres in size.

有多少人知道有的鱼会自己养殖食物？有多少人知道蚂蚁和白蚁在 3 亿年前就已经开始耕种了？列出所有你能找到的自己种植食物的动物。与你的朋友和家人分享你的清单。仔细观察他们得知后有多么惊讶。分享你清单上一些物种的更多信息，特别是小热带鱼，以及它们如何有效利用 0.6 平方米的空间度过一生。

学科知识
Academic Knowledge

生物学	表层藻基质；培养共生；草皮海藻；体外受精；小热带鱼与小丑鱼关系密切，小丑鱼也是雀鲷科的一员；白蚁和蚂蚁需要纤维来种植蘑菇，小热带鱼依靠纤维来清除海藻。
化 学	小热带鱼没有必要的酶来消化海藻纤维；用化学方法控制不想要的植物基本上就是焚烧这些植物；天然除草剂的作用；残余活性可以使地面几乎永久荒芜；醋和d-柠檬烯类除草剂；草甘膦阻碍了合成蛋白质所需的氨基酸的合成；除草剂阻断了植物细胞膜所需脂质的产生；除草剂是配制成盐或酯的酸；作为佐剂的表面活性剂。
物 理	环嗪酮除草剂抑制植物体内电子的转移从而抑制光合作用，导致植物枯萎；除草剂是水溶性的、可乳化的或悬浮在水中的颗粒。
工程学	植物除草剂的代谢工程或酶改性；作物除草剂耐受性的基因工程。
经济学	通过为有益植物创造空间来提高种植效率的重要性。
伦理学	有些人为了喂饱自己而发动进攻，而有些人是为了保卫家园；在马来亚和越南战争中，除草剂被用作化学武器，造成了超过一百万例出生缺陷；除草剂使用的增加导致帕金森病发病率上升；除草剂对鸟类和青蛙的数量有负面影响。
历 史	1940年发明的第一个合成除草剂为2.4-D；草甘膦发明于1974年，用于非选择性杂草控制；早在人类出现之前，动物就已经除草数百万年了；碧雅翠丝·波特写的彼得兔是很受欢迎的儿童故事。
地 理	在墨西哥恰帕斯古新世岩层中发现了最古老的小热带鱼化石。
数 学	海胆径向对称分为五等分；除草剂毒性对作物生长和产量影响的数学模型。
生活方式	在自己的花园里种菜可以恢复体力，放松身心；把工作变成你最喜欢的活动。
社会学	不同个体之间的合作创造了联结；当我们注意到强度或压力，并产生共鸣时，我们会分担担忧。
心理学	求偶行为；紧张的工作可能导致压力；通过问一系列问题来参与谈话；与侵略者相比，弱小无力仍决定保卫自己领土的决心；通过宣布或重复警告即将发动的攻击来威慑；勇敢。
系统论	清除不需要的物种，提高有益藻类的生长效率，甚至当藻类被吃掉，它们也能茂盛生长；蚂蚁和白蚁在有限的密闭空间里耕种，小热带鱼在开放的空间里耕种；抗除草剂。

情感智慧
Emotional Intelligence

海　胆

海胆对小热带鱼的行为感到好奇。他问了很多问题，并对答复表示惊讶。然后他分享了自己的感想。他执着于不停地提问，并不关心小热带鱼如何接受他的意见。他将小热带鱼的耕种方式与人类的进行比较，又与蚂蚁和白蚁的进行比较。当他表示喜欢吃养殖的红藻时，就从单纯的询问变成了威胁。不过当小热带鱼表现出抗争和决心时，他意识到小热带鱼有良好的防御系统。海胆改变了对海洋中哪些物种最具攻击性的看法。

小热带

小热带鱼很有信心。即使海胆可能会偷她的食物，她也有与其正面抗争的勇气。海胆不了解她为什么要照料藻类花园，她对此很失望。她耐心坦诚地回答海胆提出的问题，包括她饮食方面的弱点。她知道海胆就是破坏花园的彼得兔，在他暗示想吃她的海藻时给出了明确警告。她展现自信，并透露她与海葵的关系，但显然高估了自己的力量。她明确表示她的攻击行为并不是为了杀害别人获取食物，而是为了保护她耕种的食物。

艺术
The Arts

小热带鱼有着最美丽的色彩，是你练习水彩画技巧的理想对象。首先，勾画出小热带鱼的简单线条，画出它特有的长长的吻部。然后准备好画笔，开始涂绘各种颜色。记得在幼鱼身上添加斑点，这些斑点通常会在它们成熟后消失。画出至少十种不同的小热带鱼，如果你还想进一步尝试，再画上一些海胆和海葵。在上色前先画一些简单的线条会容易些。与你朋友的画比较一下，看看哪个小热带鱼最有攻击性。

思维拓展
Systems: Making the Connections

农业被视为人类独有。从狩猎采集到驯养动植物的进化提供了恒定食物。这导致了我们生活方式和社区组织结构的转变，促进了文化的发展。但我们很少关注到其他物种也在实现从捕食到保护食物的转变。

我们为生产食物和养育后代开拓领地。小热带鱼的行为和人类非常相似，这种鱼能在狭小的空间内满足所有需求。韧性和效率是其生活方式的重要组成部分。这给人类提供了一个有趣的案例，现代经济学显然选择了绝对优先考虑效率，并不惜一切代价追求更高的产出，即使这意味着韧性降低。

小热带鱼在约0.6平方米的区域里就能满足一切所需。最重要的是，它改变饮食习惯以适应其脆弱的消化系统——无法消化构成海藻的纤维成分。小热带鱼与藻类、以细菌为食的微生物以及海葵建立起互惠的群落。

小热带鱼有300多种，野生环境下可存活9到10年，对于体重不到10克的小鱼来说是惊人的长寿。太平洋等主要海洋中都能见到它们的身影，它们已成为海洋"农夫"。

动手能力
Capacity to Implement

你能想象藻类养殖吗?我们知道如何从种子培育植物，也知道如何饲养动物，但我们能不能尝试建立一个小型的藻类农场呢? 这比你想象的要容易。你可以在盐水或海水中种植藻类，只是一开始你要确保水是清澈的，并且有足够的阳光。你的一屋子容器里将会长满藻类! 如果任其生长，很可能会变成黏糊糊的。这并不是说这些是黏糊糊的藻类，而是生物膜上的细菌享受着与它们一起生长的乐趣。

故事灵感来自

This Fable Is Inspired by

金伯利·佩顿
Kimberley Peyton

金伯利·佩顿在佛罗里达理工学院获得理学硕士学位，目前在夏威夷大学马诺阿分校攻读博士学位。她之前的研究主要集中在（特克斯和凯科斯群岛的）水产养殖，还有无脊椎动物胚胎及其共生藻类（在佛罗里达的印第安河泻湖）。她的论文主要研究入侵藻类对夏威夷软沉积物群落（主要是原生海草床）的影响。其他研究兴趣包括在关岛发现的海草的生态学。金伯利是一名训练有素的潜水研究员，拥有高级潜水员、技术减压潜水等证书。

图书在版编目（CIP）数据

冈特生态童书.第七辑：全36册：汉英对照 /
（比）冈特·鲍利著；（哥伦）凯瑟琳娜·巴赫绘；
何家振等译.—上海：上海远东出版社，2020
ISBN 978-7-5476-1671-0

Ⅰ.①冈… Ⅱ.①冈… ②凯… ③何… Ⅲ.①生态
环境–环境保护–儿童读物—汉英 Ⅳ.①X171.1–49

中国版本图书馆CIP数据核字（2020）第236911号

策　　划　张　蓉
责任编辑　程云琦
助理编辑　刘思敏
封面设计　魏　来李　廉

冈特生态童书
耕种鱼食

[比]冈特·鲍利　著
[哥伦]凯瑟琳娜·巴赫　绘

颜莹莹　译

记得要和身边的小朋友分享环保知识哦！
八喜冰淇淋祝你成为环保小使者！